SWEAR WORD COLORING BOOK

CUPCAKES

Cupcakes and Flowers Mandalas Designs

Bullshit

VOL.2

COLOR TEST PAGE

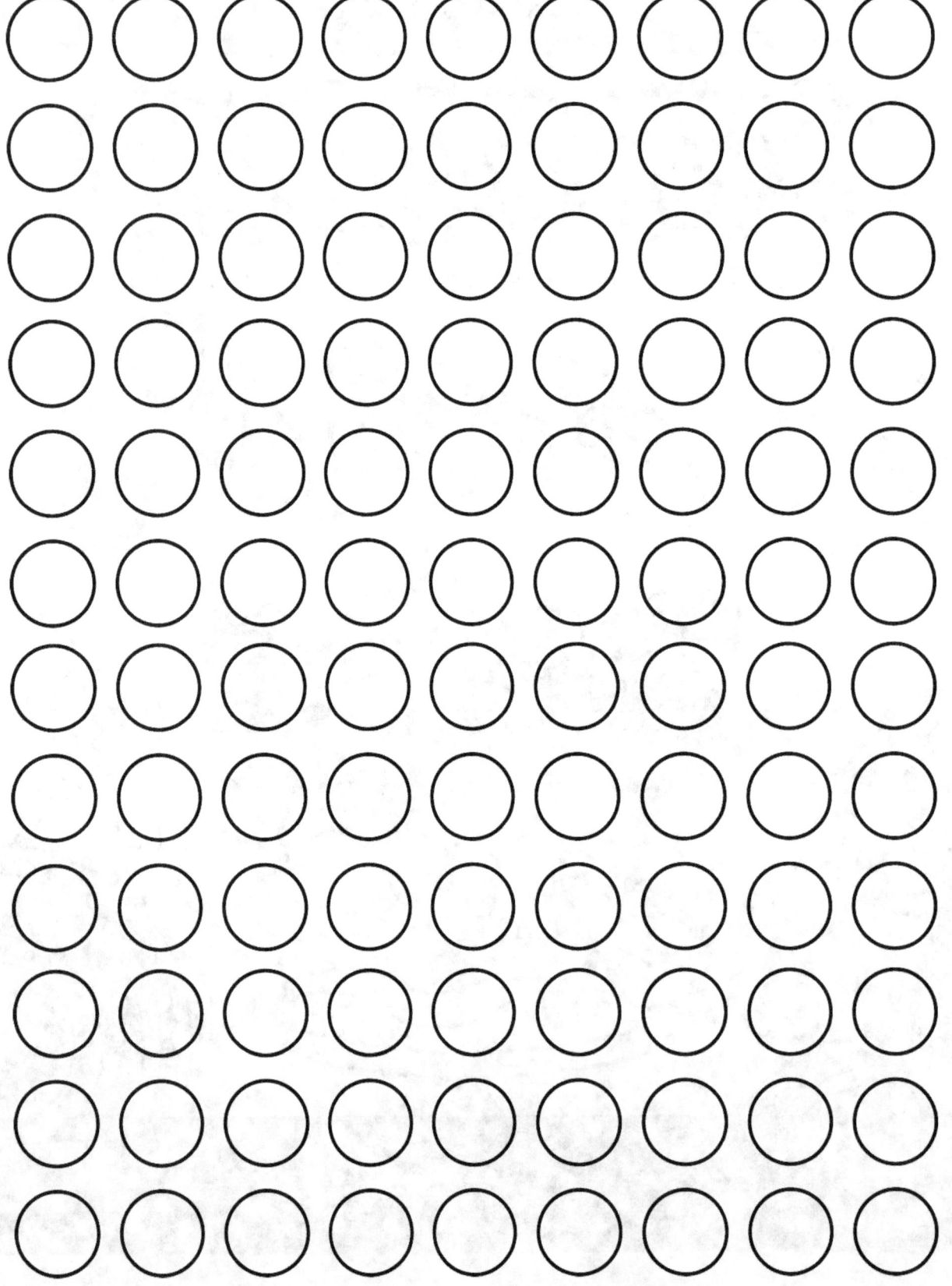

COLOR TEST PAGE

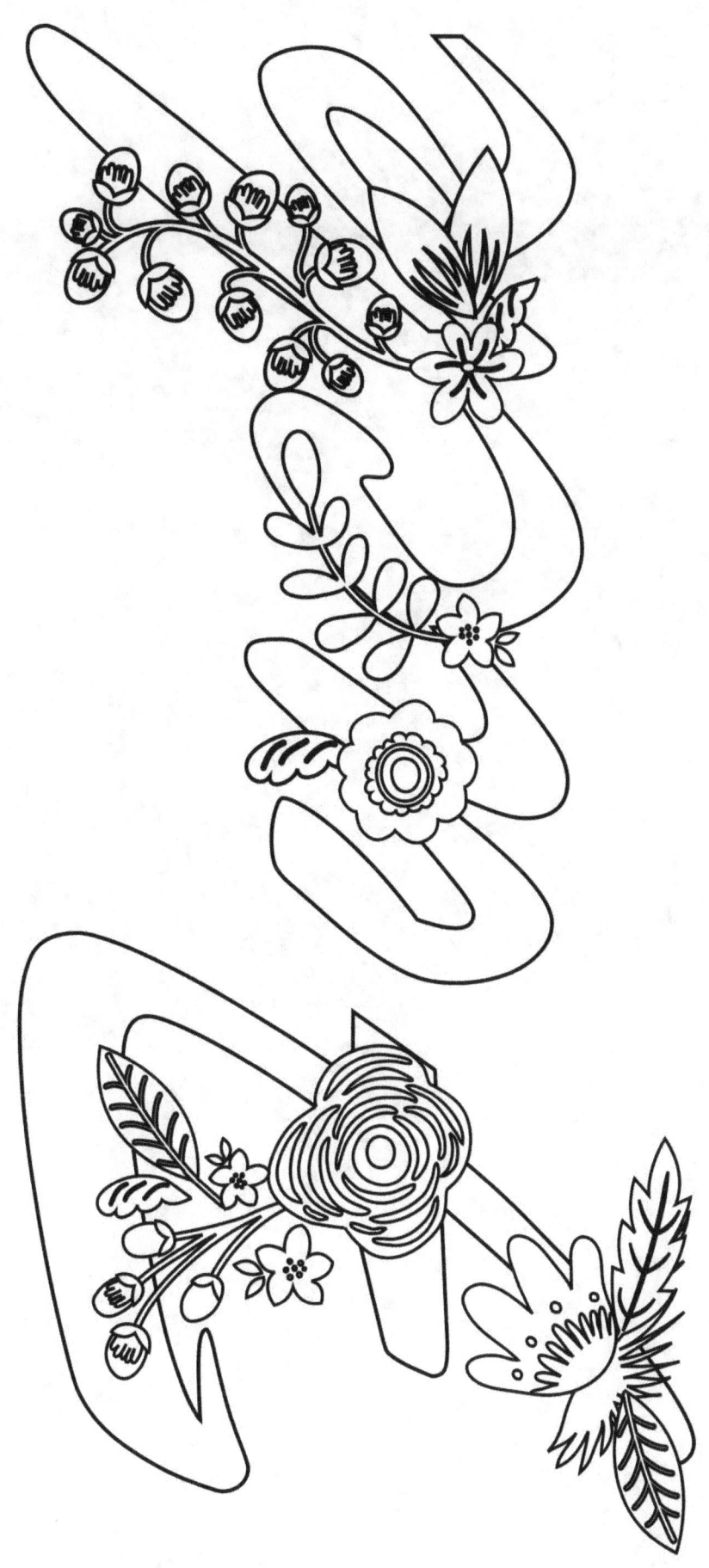

Sugartits

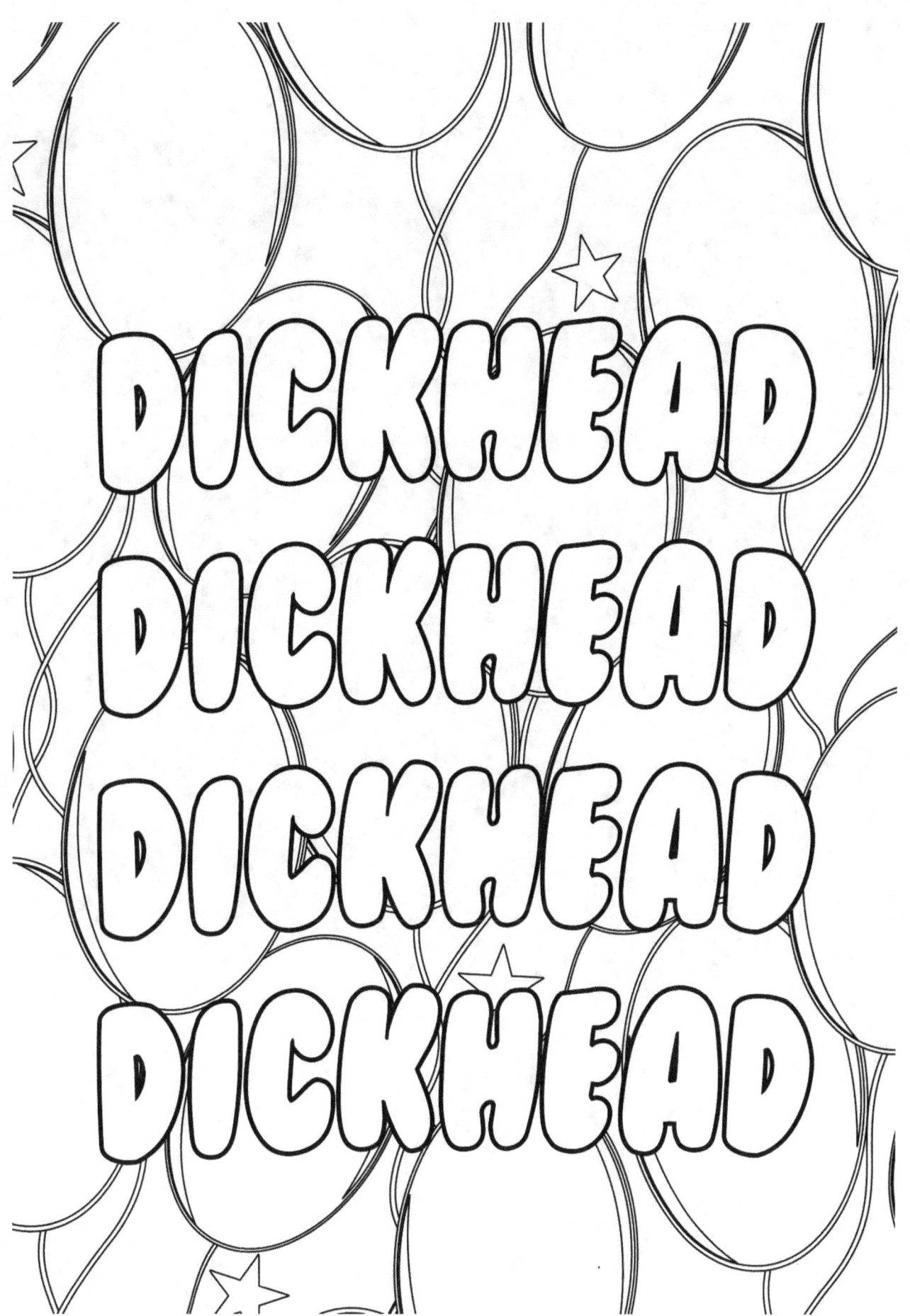

www.ingramcontent.com/pod-product-compliance
Lightning Source LLC
Chambersburg PA
CBHW081259180526
45170CB00007B/2496